Professor Noah's Spaceship

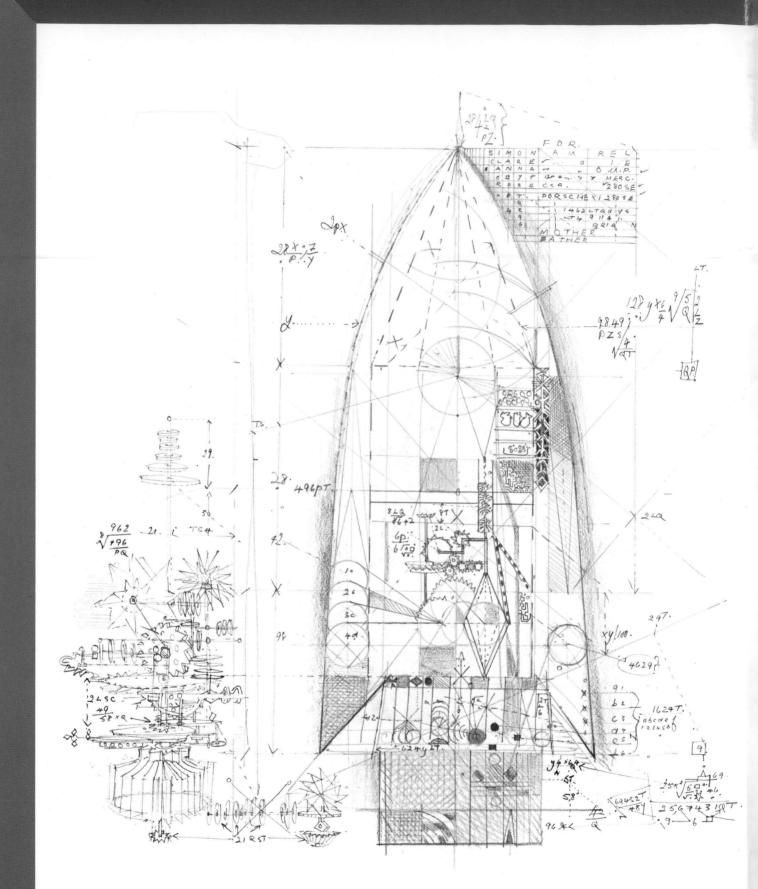

Professor Noah's Spaceship

BRIAN WILDSMITH

Oxford University Press

OXFORD TORONTO MELBOURNE

Once upon a time there was a huge forest,

and all kinds of animals and birds lived happily there.

As the years passed,
the forest began
to change.
The air around it
started to smell
and turned a nasty
colour.
No longer did the
sunlight filter
joyfully through
the leaves.
The leaves then
started to fall
from the trees,
and their fruit
began to turn bad.

A strange sadness came over the forest.
The animals were bewildered.
'What's happening?' they cried.
They became so desperate that Lion,
the king of beasts, decided to call
a meeting of all animals and birds.
'What's happening to our forest?'
they all cried.
'It is clear that we are in
grave danger,' roared Lion.
'Our forest is being destroyed,
and very soon we shall have
no place to live in.'

'The air is so foul that when I run fast I get out
of breath,' growled Cheetah.
'The bananas aren't fit to eat,' chattered Monkey.
'I'm hunted for my fur,' howled Coati.
'The oranges are terrible,' bellowed Elephant.

'When I sit on my eggs to hatch them, they break,'
piped Pelican.
Lion sighed. 'My friends, we must do something.
Our very lives are in peril.
Does anyone have an idea what we should do?'

'Owl, you are the wise one amongst us,' said Lion.
'What must we do?'
'During my flights over the forest,' said Owl,
'I have observed a huge and wondrous object being built.
Whoever is building this must be very clever.
He should be able to tell us what to do.
We must go there at once. Follow me.'

They arrived at a huge fence surrounding
a strange object.
'I must see what is going on,' said Lion.
'Elephant, lift me up so that I can see
over the fence.' And so, Elephant lifted him up.
'Extraordinary! My, my, how very extraordinary!'
said Lion. 'I have never seen anything like it.
Whoever is building this must be very clever indeed.
We must go in and see him.'

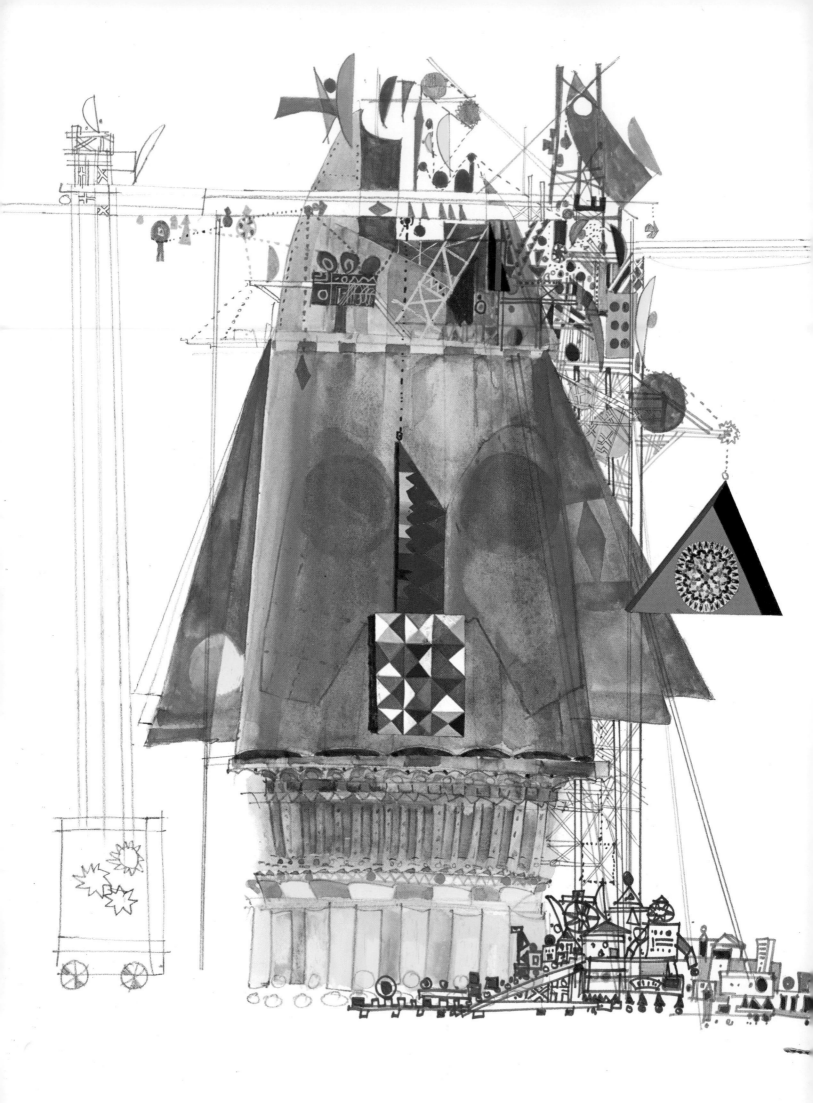

They banged on the door.
It opened silently, and in they trooped.
They all looked in awe at the amazing object.
A man came forward to greet them.
'Hello,' he said. 'My name is Professor Noah.
Can I help you?'
So Lion explained the problem to him.

'My friends,' said Professor Noah, 'it is because of
what you tell me that I am building this spaceship.
It can fly like a bird, but it will go much faster
and very much higher. It can fly to the moon
or it can fly to the stars. In fact, it is going
to fly to another planet, another world where
the forests will be different. But they will be as
beautiful as our forest once was before it was
spoiled by pollution.
Would you like to come with me?'
'Oh, yes,' they all said. 'How exciting it will be.'

'The spaceship will soon be finished,'
said Professor Noah. 'My robots do most of the work,
but I could do with a little help.'
And so the animals helped, as much as they could.
When the spaceship was nearly finished, they all
had a very good time, playing with the robots.

One day Professor Noah announced,
'My friends, we have not much time left.
There is still work to be done.
We shall need food for each of you,
as our journey will last forty days and forty nights.'

When the work was almost finished, Owl came flying in from one of his trips round the forest, and cried in fear, 'The forest is on fire. They are not only polluting it— now they are burning it.'

'Hurry,' cried Professor Noah.
'We must finish the work quickly.'
When all was ready, they boarded the spaceship.

Professor Noah called out, 'Take your positions for blast-off!'
The great spaceship launched into the heavenly skies with a mighty roar, just before the flames reached it.
Up and up, faster and faster they soared.
Looking back, the animals saw earth getting smaller and smaller, until it looked no larger than a shiny button.
On and on they went into space. Suddenly, the great spaceship waggled from side to side.

'Oh dear,'
said Professor Noah.
'One of our time guidance
fins has been damaged
on take-off. Soon we must
travel through a time-zone
which will take us
hundreds of years into
the future and help us
reach our new planet.
Our time-zone guidance
fin must be in the
correct position.
I need a strong volunteer
to go into space and twist
the fin back into shape.'
'I'll go,' said Elephant.
He put on a special
space-suit, went out
through the air-lock,
and pulled the fin into shape.
'Hurray!' they all shouted.
Professor Noah gave him
ten oranges and called
him a hero.

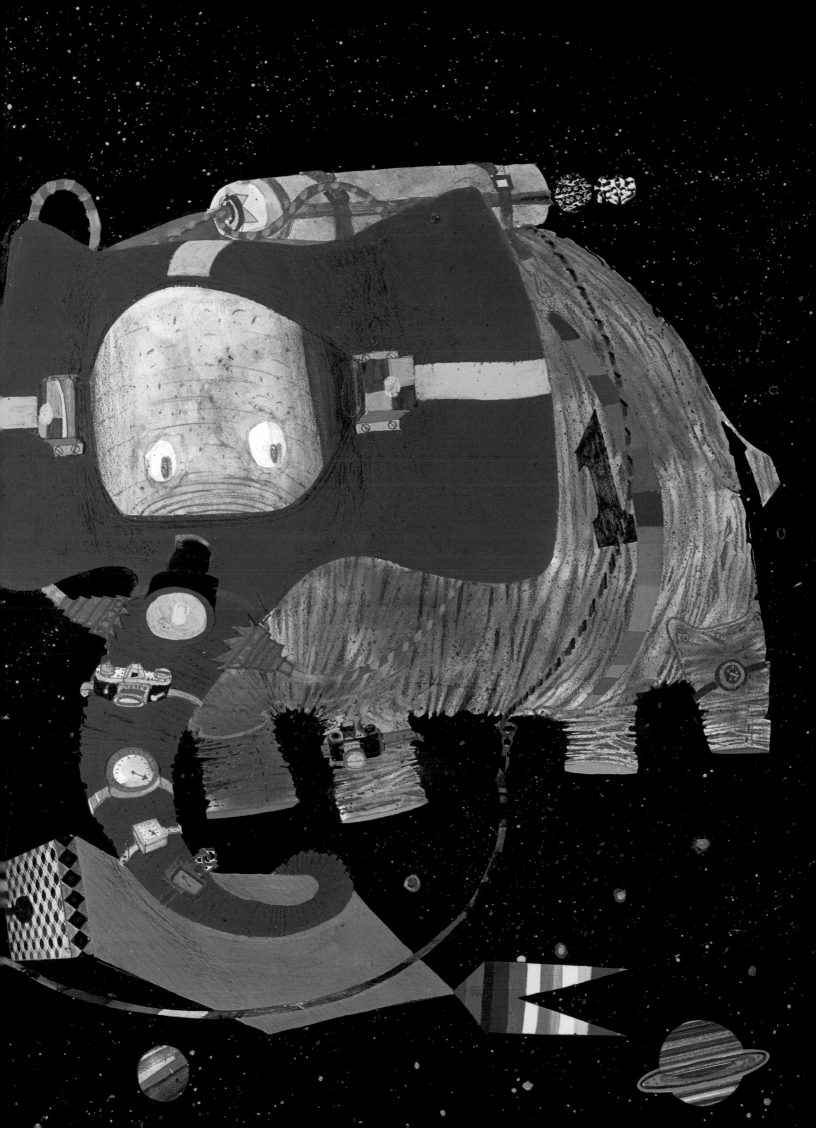

For forty days and forty nights
they travelled through space.
The further they went,
the sadder the animals became.
'Oh dear, I'm so homesick,' said Gnu.
'Me, too,' cried Llama.
'We're all homesick,' everyone wailed.
'Oh dear, oh dear,'
said Professor Noah.

At last they approached their
new planet. 'I must make sure
all is well for us there,'
said Professor Noah.
'Dove, go out and bring me
back a leaf from a tree.'
Dove did as he was told
and came back with a leaf.

Professor Noah tested it in his computer.
'Why, I can't believe it,' he cried. 'It is most
extraordinary. This is a leaf from planet Earth.
I must check my calculations.'
After making many checks, he called a meeting.
'My friends,' he said, 'when Elephant went out
to repair our time guidance system, he must
have twisted it the wrong way.

We have travelled *backwards* through time,
back to planet Earth as it was many hundreds
of years ago, before it was polluted.'
'Hurray!' they all cheered.
'Does this mean that we have come back to Earth
as it was in the beginning?' said Lion.
'Yes,' said Professor Noah. 'It is a wonderful world,
and we must keep it that way.'

As the sun rose,
the spaceship
descended
to Earth.
The animals
all thanked
Professor Noah
for saving them.
And then they left
joyfully for their
new homes.
'How lovely it is,'
they all cried.

'Yes,' said Otter, 'and thank goodness for all the rain. There seems to have been some flooding here.'

Oxford University Press, Walton Street, Oxford OX2 6DP

Oxford New York
Athens Auckland Bangkok Bogota Bombay
Buenos Aires Calcutta Cape Town Dar es Salaam
Delhi Florence Hong Kong Istanbul Karachi
Kuala Lumpur Madras Madrid Melbourne
Mexico City Nairobi Paris Singapore
Taipei Tokyo Toronto

and associated companies in
Berlin Ibadan

Oxford is a trade mark of Oxford University Press

Copyright © Brian Wildsmith 1980   ISBN 0 19 272149 6

First Published 1980 as Hardback
Reprinted 1985, 1986, 1987, 1989 (twice), 1991, 1992
First Published 1985 as Paperback
Reprinted 1986, 1987, 1989, 1991, 1992, 1996

British Library Cat. Pub. Data

Wildsmith, Brian
 I. Title.
 823'.9 1J   P27.W647   80-40984
 ISBN 0−19−279741−7

Printed in Hong Kong